WROUGHT IRONWORK

FOX CHAPEL
PUBLISHING

WROUGHT IRON GATE. Probably made by Warren about 1720. It stands at the entrance to Clandon Park, seat of the Earl of Onslow.

WROUGHT
IRONWORK

*A manual of instruction
for craftsmen*

Published by
RURAL DEVELOPMENT COMMISSION
SALISBURY

Dear Reader:

We're very happy to bring the classic *Blacksmithing Series* from COSIRA (Council for Small Industries in Rural Areas) back into print. In the 35 years I have worked as a bookseller and publisher, certain iconic titles stick out. This series is one of those considered a classic because the photographs, text, and illustrations are so complete and highly focused on its subject.

The series:
- *Blacksmith's Craft* (978-1-4971-0046-6)
- *Wrought Ironwork* (978-1-4971-0064-0)
- *Decorative Ironwork* (978-1-4971-0063-3)

You hold the *Wrought Ironwork* volume in your hand. As you read, please be aware that we have not made any attempt to update the techniques or tools. Consider this treasure trove of knowledge a time capsule from the past.

May you be inspired to pick up the blacksmith's hammer and try your hand at this ancient skill.

Enjoy!

Alan Giagnocavo, Publisher
Fox Chapel Publishing

ISBN 978-1-4971-0064-0

Library of Congress Cataloging-in-Publication Data

Names: Council for Small Industries in Rural Areas (Great Britain), author.
Title: Wrought ironwork / Council for Small Industries in Rural Areas.
Description: Mount Joy : Fox Chapel Publishing, 2019.
Identifiers: LCCN 2019018636 (print) | LCCN 2019020629 (ebook) | ISBN
  9781607657323 (ebook) | ISBN 9781497100640
Subjects: LCSH: Blacksmithing.
Classification: LCC TT220 (ebook) | LCC TT220 .W76 2019 (print) | DDC
  682--dc23
LC record available at https://lccn.loc.gov/2019018636

To learn more about the other great books from Fox Chapel Publishing, or to find a retailer near you, call toll-free 800-457-9112 or visit us at *www.FoxChapelPublishing.com*.

We are always looking for talented authors. To submit an idea, please send a brief inquiry to acquisitions@foxchapelpublishing.com.

Printed in China
First printing

Because working with metal and other materials inherently includes the risk of injury and damage, this book cannot guarantee that creating the projects in this book is safe for everyone. For this reason, this book is sold without warranties or guarantees of any kind, expressed or implied, and the publisher and the author disclaim any liability for any injuries, losses, or damages caused in any way by the content of this book or the reader's use of the tools needed to complete the projects presented here. The publisher and the author urge all readers to thoroughly review each project and to understand the use of all tools before beginning any project.

CONTENTS

CHAPTER 4   MAKING AN ORNAMENTAL GATE—*continued*

LESSON

CHAPTER 5   FITTING AND ASSEMBLING THE GATE   83

LESSON

PART III

PREFACE

In recent years, wrought ironwork has regained some of its previous popularity and it seems likely that the severity of our modern buildings may be relieved by this traditional form of decoration.

Orders for wrought ironwork are welcomed by many rural blacksmiths, not only for the income they bring, but as a pleasant change from the daily routine of an agricultural smithy. Some smiths are, however, out of practice and lack confidence in their skill. So this book has been prepared by the Rural Development Commission, which provides a national advisory service for rural craftsmen, to help them to refresh their technical knowledge and to provide apprentices with a basic introduction to this subject. It will also supplement the practical instruction which the Commission gives to rural craftsmen in their own workshops.

Detailed advice on design, which is a most important aspect of the craft is not given here; but a high degree of technical skill is of no avail if a sense of design is lacking. This can be developed by taking every opportunity to see fine examples of traditional and contemporary wrought ironwork, and by supplementing this with a careful study of the books which are listed on page 96. The Commission publishes a Catalogue of Drawings for Wrought Ironwork which is sold to the public, although the library of the working drawings is only available to rural craftsmen.

The system of describing techniques by sequences of still photographs, briefly captioned, proved very successful in The Blacksmith's Craft and has been used again in this book. Where methods vary, the one most suitable for the beginner has been described.

INTRODUCTION

METAL WORKED ON THE ANVIL has a grace which belies its strength, and is particularly suited to gates, railings, grilles, sign and lamp brackets, as well as such hearth furniture as fire-dogs, screens, pokers and tongs.

The first part of the book describes the making of the most common decorative features such as scrolls, water leaves, wavy bars and twists. The difficult acanthus leaves and embossed work are not included in the present volume. Part 2 describes step by step the making and assembly of a gate, which includes the same techniques as are used in all traditional decorative ironwork such as grilles, brackets and hearth furniture. The final chapter discusses the problems of painting and rust proofing. A knowledge of basic smithing techniques has been assumed throughout, and only those tools peculiar to decorative ironwork are mentioned. Basic smithing techniques, heats and tools are described in *The Blacksmith's Craft.** Craftsmen are also urged to study the books on book-keeping; costing, estimating and business methods listed on page 96.

The techniques are shown by sequences of photographs with concise explanations which are intended to supplement instruction on the anvil. It is hoped that this clear and practical method will help the smith to achieve the highest standard of work.

TOOLS

As well as the tools and equipment which every blacksmith keeps in his shop, some special tools are used for decorative ironwork. For scroll-work you will need wrenches, pliers and scroll tools. Three scroll wrenches are shown in Fig. 1: several sizes are necessary to suit the shape and width of different scrolls.

Fig. 1

The Blacksmith's Craft published by the Rural Development Commission.

Fig. 2

Fig. 3

Fig. 4

Fig. 5

Round-nosed pliers (Fig. 2) are used for gripping the tip of the scroll to the scroll tool, for adjusting the curve of a scroll nearly completed, and for fitting collars.

Bow pliers (Fig. 3) are also used for fitting collars: they are designed not to spoil the collars when they are gripped.

When several scrolls are required to the same design, it is usually worth while making a scroll tool, or adapting one already made: making a scroll tool is described on page 34.

The halfpenny snub-end scroll tool (Fig. 4). The groove is made with a one-inch fuller. The top should be curved as shown and the edge sharp. This tool is shown in use on page 24.

10

Fig. 6

Fig. 7

The monkey tool, side set and butcher are used in making a shoulder. The monkey tool (Fig. 5) is used for squaring the shoulders of round tenons. If the end of the tenon is in danger of fouling the base of the hole in the monkey tool, it can be seen through the cross-hole and shortened.

The side set (Fig. 6) is for squaring the shoulder precisely; the end is bevelled to an angle of about 75°. A butcher (Fig. 7) is useful for making shoulders quickly.

Fig. 8

For leaf work you will need a leaf hammer and a leaf tool (Fig. 8) and a crimping tool. A leaf tool is simply a forked stake with the inner edges slightly rounded so as not to gall the leaf. The head of the crimp tool (Fig. 9) is hollowed and rounded for crimping the leaves.

Fig. 9

Fig. 10   *Fig. 11*

Lastly, for measuring off and marking, you will need a smith's square (Fig. 10), dividers, and chalk (Fig. 11) – both engineers' chalk for marking metal, and schoolroom chalk for transferring drawings on to metal plate.

To Transfer a Drawing on to a Metal Plate

Whenever metal has to be shaped hot to a drawing, the drawing must be transferred to an iron plate. A sheet of brown paper, chalked on one side is used like carbon paper to transfer the drawing on to the plate.

Fig. 12

A

First choose a piece of plate with a surface neither too new nor too rusty. Spread out the brown paper, and file a piece of natural or school chalk on to it with a bastard file.

12

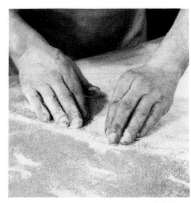

Fig. 13

B

Rub the chalk in with your finger tips.

Fig. 14

C

Blow off the surplus chalk, turn the paper over, lay the drawing on top, and go over the lines with a pencil.

Fig. 15

D

This is the result. Strengthen any weak lines with engineers' chalk.

PART I

CHAPTER 1

MAKING AND FIXING SCROLLS

Scrolls are used more often than any other motif in wrought ironwork. Their variety and grace enhances many of the finest designs; a mastery of scroll-work is, therefore, the first essential. The beauty of a scroll depends upon its proportions. As a smith gains experience, he will train his eye to judge how heavy a section of metal to choose, the fineness of the taper and how tightly to roll the scroll, and he will train his hand to strike the hammer blows where they tell. There are no rules by which to determine the proportion of a scroll and even when working to drawings, a smith must use his own critical sense to ensure that each scroll has a graceful flow.

Practice in freehand drawing is a valuable training. A beginner should take a course in drawing if he can. If there is no school near enough to attend, then he should practise drawing a simple scroll true to line when he has a few moments to spare. At first he may find it surprisingly difficult, but if he perseveres he will find the training most useful when he comes to make a scroll in metal.

This chapter describes nine different types of scroll, beginning with a ribbon-end scroll, which is the simplest, and progressing to the bevelled scroll. The scrolls have been explained before the scroll tool, because the making of the tool is best left until the smith has some experience of its use. The chapter ends with scrolls worked into a 'C' and an 'S', and the fixing of scrolls with collars. The side panels and centre panels for a gate described in Part II are typical examples of decorative scroll-work.

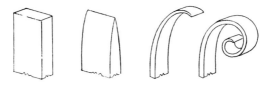

Fig. 16

A

The ribbon end is drawn down in two stages.

First, on the face of the anvil, reduce the width of the bar. Do not allow the thickness to increase, but do not reduce it either.

B

Next, move on to the bick, turn the metal on to the flat and draw the taper down.

The reduced width now spreads to the original size or a little over.

C

With the bar on edge, round up the tip neatly.

D

Level up on the edge to the width of the parent metal. If the first stage was well judged, little will have to be done now.

E

Roll the tip of the scroll over the edge of the anvil. Start at the extreme end and take care not to chop the metal against the corner of the anvil.

F

Now continue to roll up the scroll, at a RED heat, on the anvil face. As the scroll is formed, both the bar and the direction in which the blows are struck should steadily approach the vertical.

G

You will probably have to repeat these movements two or three times.

16

H

Scrolls can be completed by this means. But they are more often finished on a scroll tool as shown next.

J

The making of a scroll tool is described on page 34. It is better to begin on one made by an experienced man and see how it works before making your own. Take a RED heat on the bar. Place the tip of the scroll which you have forged on the tip of the scroll tool. Hold them together with round-nosed pliers. Pull the end round far enough to ensure that the end of the scroll has firmly gripped the tool.

K

Relax your grip, and lower the scroll from the raised tip of the tool to the level of the main part.

L

By now the cold end of the scroll will grip the scroll tool by itself. Continue forming the scroll, forcing the metal close to the scroll tool with a scroll wrench. Very small scrolls can be bent cold by this means.

M

Move round the anvil as the scroll is formed so as to work in a comfortable position.

N

Continue in this way until the scroll is finished. It is sometimes a good idea to mark the scroll tool with chalk to show when the scroll is the right size.

O

A well made scroll tool will produce a scroll which is almost flat, but the scroll will require a little trueing up at the finish.

FISHTAIL-END SCROLL

P

Fishtail-end Scroll
The fishtail-end scroll is similar to the ribbon-end scroll except that instead of being kept parallel, the metal is spread out as it is forged.

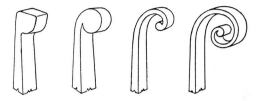

Fig. 17

A

The snub-end is formed over the square edge at the far side of the anvil, or, if the anvil is worn, over a square-edged stake as shown here.

The amount of metal projecting over the edge should be equal to the thickness of the bar.

B

Forge in the swelling and level up on edge.

C

In order that a solid snub-end scroll shall look graceful, the metal should be forged to less than half its original thickness for a considerable way behind the snub, scarcely widening at all.

D

Forge in the outside corner of the snub over the anvil or stake edge.

E

Turn the back of the piece on to the anvil face and forge in the remaining corner.

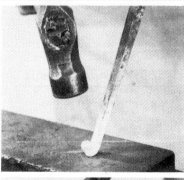

F

Dress up the snub, which should be perfectly round. It should join the taper with a flowing curve so that the scroll may bend gracefully from the snub.

G

When making any kind of snub-ended scroll on a scroll tool, it is vitally important to give the scroll a good start with hammer and pliers (as in Lessons 1 and 4).

The curve of the scroll must fit the scroll tool and grip it. On no account must the snub be allowed to do any of the holding or it will be distorted; this is very difficult to rectify.

Here is the finished scroll.

20

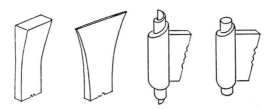

Fig. 18

A

Reduce the metal on the bick of the anvil, leaving an untouched lump at the end and a long taper. It is shown here on edge, cold.

In the same heat, move it on to the anvil face, turn it on edge, and level it up.

B

Take a WELDING heat on the end in case the metal has cracked, and forge the lump into a broad, short fishtail.

C

Notch the end as shown. This, by removing surplus metal from the centre, makes it easier to roll a tight snub, and prevents the middle of the snub from bulging when the end faces are burred.

D

Starting at the extreme tip, begin to roll up the snub end.

E

Tighten the roll in the same way as the solid snub was formed.

F

The projecting ends can be tightened up neatly with the round-nosed pliers. A scroll is sometimes left in this form, and is called a Fishtail Knib. It is delicate in appearance but apt to catch on clothing, so it is more often finished in a Fishtail Snub as below.

G

Flatten the end faces, burring them carefully so that the ends of the roll appear solid. If one end flattens more readily under the hammer than the other, cool it a little with water.

Fig. 19

A

As in making a solid snub-end (see page 19) it may be wise to use a square-edged stake.

With the metal projecting as far over the edge of the anvil or stake as the bar is wide, forge an offset neck.

B

So far, the snub is being formed in the same plane as the bar, but in the finished scroll it must be at right angles to this. The snub will, therefore, need to be twisted through a right angle. Before making this twist, round up the neck, as the corners would show up the twist and it is more difficult to forge them in afterwards.

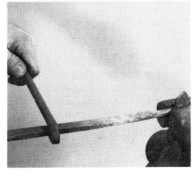

C

Take a BLOOD RED heat, and concentrate it on the rounded part, using water if necessary.

Now grip the snub in the vice and twist the bar at right angles using a suitable wrench or a pair of tongs.

D

Take a heat on the snub and forge in the far corner.

E

The neck which was rounded for twisting should now be squared up again. If this were left till later, the head would get in the way.

F

Turn the piece on its back and round up the snub on the anvil face.

G

Here is the special tool (shown on page 10, Fig. 4) with the piece, showing the radius in the neck as forged so far.

H

To finish the snub and make it blend gracefully into the neck, take a near welding heat, cool the far edge quickly to avoid damaging it and get rid of the radius with light blows.

J

As the normal scroll tool cannot readily be adapted to these scrolls, it is usually easier to bend them with the round-nosed pliers.

If a large number is needed, a special scroll tool with an open centre should be made.

Lesson 5 **BOLT-END SCROLL**

Fig. 20

A

Forge a fishtail, leaving the end not less than ⅛-inch thick. If the bolt-end is to be wide in proportion to the bar, it will be necessary to upset the end to some extent first.

25

B

The bolt is made from a round bar, and should be at least ¼ in. longer than the width of the fishtail.

Make a notch all round the bar with a hot set, leaving just enough metal at the centre to support the bolt while it is being welded.

Hold the set to one side, as shown here, so as to leave a square end on the bolt.

C

Help will be needed for this weld. Take a WELDING heat on both pieces. Have your mate take a wire brush in one hand and the round bar in the other.

He should lift the round bar from the fire and place it in a bottom swage, wire brushing it on the way.

Lift out the fishtail and present the end, bottom upwards, to your mate. He should give it one stroke with the wire brush. Turn it over and immediately weld it to the bolt.

D

Hand your hammer to your mate and sever the bolt from the bar with a hot set.

E

Square up the ends of the bolt.

F

Very slowly, so as not to burn the thin fishtail, take a FULL WELDING heat and strengthen the weld, rolling the bolt up a little.

G

Grip the bolt in a vice and pull the bar over to begin the scroll.

H

Take a fresh grip and repeat the process until the bar almost encircles the bolt, then form the rest of the scroll with hammer, horns and wrench (see page 32 E).

Bolt-ended scrolls are not very common and are normally fairly sturdy, so scroll tools are seldom used for them.

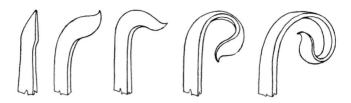

Fig. 21

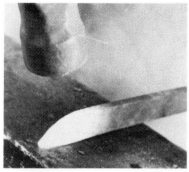

A

Forge the end of a bar to a radius. The curve should be no longer than shown here, otherwise the leaf-end will be too long.

B

Neck over the bick, leaving the straight bottom edge of the end a little longer than the width of the bar.

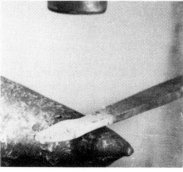

C

Thin the neck on the flat, to about two-thirds.

D

Bend the neck, on edge, over the bick.

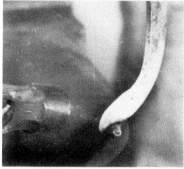

E

Cock the tip, still on edge, in the opposite direction to the main bend.

F

With ball-faced hammer, thin both edges, on one face, controlling the curve as you do so.

G

So far, the piece has been bent on the edge. Now, with the hammered face upwards, curve the leaf on the flat.

H

The tip has already been cocked in the opposite direction to the main curve on edge (E). Now cock it in the opposite direction to the curve you have just made.

J

Increase the curl in the neck at the same time twisting the leaf partly into line with the bar. Avoid solid blows which would distort the shape.

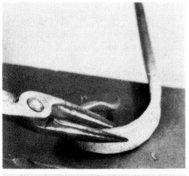

K

Form and adjust the scroll with pliers.

L

The finished scroll. The leaf can be aligned a trifle if necessary, with light blows.

Fig. 22

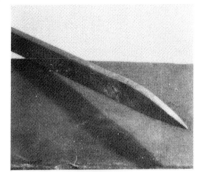

A

Draw the end of the bar with one edge curved, the other straight, and the extreme tip sharply pointed. Curl on bick.

B

With ball-faced hammer, thin both edges on one face, controlling the curve as you do so.

C

Tap the end of the scroll out of alignment and tuck in the end, neatly, on the anvil face.

D

Here is the result so far, shown cold. Note the steady increase in radius from the tip.

E

Place the inner edge of the tip on the point of the bick, the outer edge being kept a little off it.

Hammer the outer edge lightly down onto the bick, bending it only. Avoid solid blows which would distort the shape.

F

Continue in the same way round the scroll, re-heating as necessary.

G

Here is the result, almost completed.

H

Complete the bending with horns and pliers.

J

By varying the position of the scroll and the pliers the scroll is not only bent but also twisted to maintain the balance between the bevel and the curve.

K

Finally the centre can be flattened a little if necessary.

Fig. 23

It is not necessary to make a new scroll tool for every job. In any established shop there will be a number of scroll tools to hand which have been made or adapted to the job in hand. Sometimes the beginning of the scroll tool only is used, a chalk mark being made to show how far the scroll bar should be pulled round. New scroll tools, however, have to be made sooner or later.

It is easier to make a scroll tool direct from the drawing than it is to make one from a scroll. So if the job warrants a new scroll tool make it before you make the first scroll. Once you know how the scroll tool is used this is not difficult.

Simply make the outer edge of the scroll tool conform to the inner edge of the scroll on the drawing. The thickness of the scroll tool does not matter.

A

Take a bar somewhat heavier than the scroll to be made.

Forge a fishtail on the end.

B

Offset the fishtail by straightening one edge.

C

Cut off the end square with the straight edge.

D

If you are right handed, begin rolling the scroll tool with the offset edge to the left, over the far edge of the anvil, or vice versa for a left-handed smith.

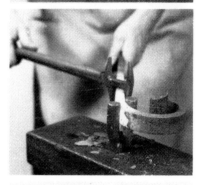

E

When you have made a good start with the hammer, continue with horns and wrench.

Pull with the wrench, rather than with the hand holding the bar, as the shape is more easily controlled by the wrench.

F

Finally bend the end of the scroll tool at right angles and wedge it into the swage hole of the anvil. The purpose of the off-set fishtail can be clearly seen; it makes the start of the scroll accessible.

The scroll tool can be held in the vice if it is more convenient.

Fig. 24

'C' scrolls are often needed in large numbers for gates, grilles, fire-screens and so on. To make them quickly it is necessary to have a scroll tool exactly the right size which finishes at precisely the centre point of the scroll. Otherwise the end of the scroll tool would foul the first scroll when the second was being made.

For measuring it is convenient to have the end of the scroll tool bent down at right angles with a fairly sharp outside corner.

First find out by measuring how much metal is needed to make each 'C' scroll, allowing for the drawing out. Next cut off all the pieces to this length and centre-punch mark them in the middle. Draw one to the correct length, chalk it, and draw all the others to it as a pattern.

Take a RED heat and scroll. The centre-punch mark should come level with the corner of the scroll tool to within ⅛ inch.

This means the scrolls will be sufficiently alike to fit together with little cold setting.

A

Here is a 'C' scroll being completed on the tool.

Fig. 25

'S' scrolls also are often used in large numbers. A scroll tool of the same type as used for 'C' scrolls is best, even though there is no danger here of the first scroll of the 'S' fouling the tool as the second is made.

Both 'C' and 'S' scrolls are used in pairs or greater numbers in repetitive designs. It is the best practice to fasten them together with collars; this is described on page 38.

However many scrolls are wanted, start by collaring them together in pairs and then fix the pairs together. To do this, first make a frame whose inside size is the over-all size of one scroll.

Offer the scroll to the frame, and adjust the scroll until it fits. Then make a second frame to fit a pair of scrolls. Press the scrolls into this frame and fix the collars.

A

The frame shown is not part of the design, but merely a jig to ensure the scrolls are accurate, and to keep them together while the initial collars are put on.

Lesson 11   COLLARS

Among the various methods of fixing scrolls together collars are important. They not only contribute to the design but are often the only satisfactory way of doing the job.

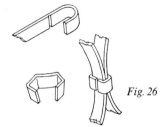

Men who are not skilled at making them often shun collars as being troublesome and expensive. But if they are made by the right methods with forethought and skill, no one need fail to master them.

Collars are frequently needed in fair numbers, so it is worth taking trouble to have all the details right before you begin to turn them out.

Fig. 26

First select or make a mandril, a piece of iron convenient to hold, the end of which is the same size as the two thicknesses of scroll which the collar will grip. The mandril can be either a plain piece of bar twice the thickness of the scroll bar, a larger piece of bar drawn down to this thickness; or, for light work, a piece of scroll bar bent back on itself, as shown in the first photograph of this lesson.

Next find out the length of the piece of metal required to make each collar. Make a trial with the actual bar from which you intend to make the collars, fitted to a pair of the actual scrolls in hand. Although measurements are needed for a start, they are not to be trusted as all bars vary slightly from their nominal size, and different qualities of iron or steel stretch a different amount when bent at the corners.

A

The measuring can be done in one of two ways.

The mandril can be laid in the collar bar as shown here and rolled along four quarter turns, and twice the thickness of the collar bar added by eye.

38

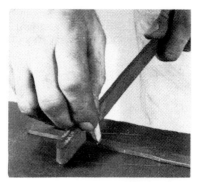

B

Alternatively a piece can be marked off to a measurement of twice the thickness plus twice the width of the mandril, plus twice the thickness of the collar bar.

Whichever method of measurement is used, set a pair of dividers to the length before making up the trial collar.

C

Notch the collar bar halfway through on a hardie.

The operations which follow are done more quickly than they can be described, in order to make best use of the heat.

D

Take a NEAR WELDING heat, and bend the collar round over the anvil bick with the cut on the inside. Take care not to draw the collar on the bick. Close up to a 'U' shape bringing the end in line with the cut.

E

Lay the mandril on the flat in the middle of the 'U' and close the ends of the collar over it. The end of the collar should meet on the side of the work, as here, and not on the front, back or corners.

F

Forge the collar clean and square on the mandril. Tap it off the mandril and level up the edges. All this can be done in one heat, but the beginner need not be ashamed of taking two. Now examine the collar and decide if any obvious alteration should be made in the length.

G

Heat the collar on the tip of the poker and open it out by putting the round-nosed pliers inside and pulling the handles apart.

H

Thread it on the scrolls.

J

Tap it down with the hammer.

K

Pinch in the sides with the bow pliers.

L

If required, line up join with pliers. Now examine the job with the greatest possible care. Decide if any alteration in the original length would be an advantage, and if so, make another trial. When finally satisfied, mark off on the metal all the pieces you will want for your collars, and notch as much at a time as you can conveniently handle.

CHAPTER 2

TWISTS, WAVY BARS AND WATER LEAVES

In this chapter are described three types of ornament which are very commonly used, and are not difficult to make. In all decorative work, a sense of design is essential if good craftsmanship is not to be spoiled by clumsy lines. If you have the opportunity, study examples of really fine work to see what an imaginative use of the qualities of iron can achieve. The ironwork at Hampton Court is famous, and the metalwork department of the Victoria and Albert Museum contains many beautiful pieces. The list of books on page 96 includes several illustrated studies of designs. As you gain experience you will learn the feel of the work, and will be able to make many interesting and varied designs within the compass of the techniques described here.

Lesson 12   **TWISTS**

Twists are second only to scrolls as a feature of decorative ironwork.

They may be made from plain square bars, flat bars, or bars forged to some special section; or bundles of bars, either the same or varied in section, may be made up and twisted. To produce a right-hand twist, turn the bar anti-clockwise and vice versa.

Square bars up to ½ inch can be twisted cold, provided that care is taken not to damage the bar where the force is applied, and that the twist is kept straight. The method is to use a wrench wide enough to twist the bar without bruising it and a piece of barrel to keep the bar straight.

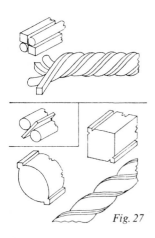

Fig. 27

A

Choose a piece of iron barrel which is slack enough on the square rod not to jam when the twist is made. Cut it to the length of the twist required. Mark each end of the twist, and grip the bar in the vice up to one of the marks. Slip the piece of barrel over the part of the bar to be twisted and apply the wrench close to the barrel end. It is convenient to support the end of the bar on a notched piece of wood, as shown here.

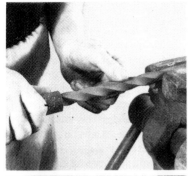

B

Here the barrel is being removed, revealing the twist which has been formed inside it.

C

The simplest composite twist is made from a bundle of four round rods welded together at the ends.

D

The finished twist is here shown cold.

E

Here, finished and wire brushed, are shown three twists.

The lower twist is made up of two square bars and two round bars. The middle one is made of a flat bar and two round bars. The top twist is a specially-formed square bar, made as shown on next page.

F

Mark off the length of the twist. On each side of two opposite corners scribe a line ⅛ inch from the edge. Mark this line clearly with a cold chisel or cold set.

Now take a heat and cut into this line with a hot set held at a slope as shown here.

This is to avoid cutting the corner right out of the bar.

G

The pair of corners which have not been cut are now rounded to an oval section.

H

Clean up now with a hot rasp or old bastard file, as it is difficult to clean up after the twist is made.

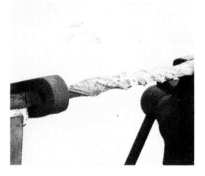

J

Now take a heat, just the length of the twist and cooling out evenly at each end. If you can do this without using water, so much the better.

Twist by any convenient means.

K

However much care is taken with the heating, the twist is very likely to form unevenly.

As soon as this becomes noticeable pour water on those parts which are twisting too much, to cool them. Sometimes a twist back is necessary. In this case the parts which are not over twisted are cooled, so that they are not affected as the bar untwists.

Lesson 13   **WAVY BARS**

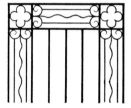

Fig. 28

Here is a wavy bar tool in use. It is quite straightforward to make, but it is important that the waves should match perfectly, because each wave in the bar, as it is formed, must mate first with one wave of the tool and then with the next.

A

It is difficult to make the first wave exactly where it is wanted, so leave a little extra metal, and trim the bar to length when the waves are formed.

Take a long heat, grip the bar to the end of the tool with a pair of tongs and begin to pull it into the first wave with a wrench.

45

B

The bar will tend to twist sideways. To counteract this, use the wrench from above and below alternatively.

C

Take another heat and grip the bar to the tool with two pairs of tongs.

D

In order to have both hands free, slip chain links over the reins of the tongs. As each new wave is made, move up one on the tool. Set the final bends with horns and wrench and trim to length.

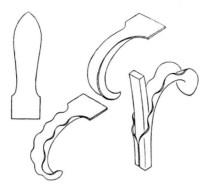

Fig. 29

Water leaves are usually made of 14 s.w.g. metal, a little under ⅛ inch in thickness.

A

Measure the length of the leaf from the drawing with a piece of string. Measure from the tip along the underside of the leaf as the metal will stretch, but is not appreciably shortened in making the leaf. This will be a rough guide to the length of metal you will need.

B

The width of the leaf plate where it is to be welded on to the bar must be the same as the distance round the bar. The widest part of the leaf is usually the same width.

Cut out a paper template. Fold this and compare it with the drawing. If there is any discrepancy make an allowance by eye in marking the metal.

C

Cut right through the metal with a cold chisel to the shape of the paper template. Use an old piece of plate or an anvil saddle to avoid spoiling the chisel's edge.

File or grind the rough edges smooth.

D

Thin out the edges of the leaf.

E

At a RED heat bend the leaf between the leaf hammer and the leaf tool to a 'U' section.

Keep the leaf straight at this stage.

F

Take a short BRIGHT RED heat and begin to curl the leaf at the tip.

G

Take another heat and extend the curl.

H

This is the result to aim at, shown cold. At this stage the curve should be greater than in the finished leaf, as the crimping will uncurl it slightly.

J

The tip of the traditional water leaf is twisted to one side or 'blown over'. The twist is worked over the end of the bick with a leaf hammer.

K

Note the number of crimps in the drawing and mark each depression on the leaf with chalk. Make a slight dent at each mark with the leaf hammer over the crimping tool (see Introduction, Fig. 9). Compare the dents on each half of the leaf to ensure that they align, then form the crimps, first from one side then the other, working them down to the flute of the leaf. Hold the leaf at an angle to the crimping tool, so that the crimps form diagonally, but always use the hammer in line with the tool.

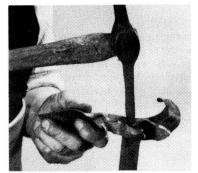

L

*Move the leaf to and fro over the tool
hammering each crimp a little in turn,
both on the outside as here –*

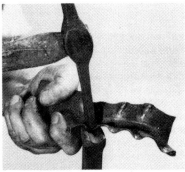

M

*– and on the inside. Near the centre of
the leaf the crimping is largely a bending action.*

N

*But towards the edge the metal is lipped
over slightly more. This emphasises the
waviness of the edge, and gives added
life to the leaf.*

O

*Grip the leaf and the bar to which you
are going to weld it, in a vice, and fold
one edge of the base over the bar. (A
distance-piece will be needed to fold the
second edge.)*

P

Now grip leaf and bar by the edges and fold the base of the leaf right round. If the leaf has been cut as in B of this lesson there will be a gap between the folded edges of the leaf equal to the amount taken up by the corners. This is correct, as the metal will stretch in welding. The bar should be slightly tapered so that the leaf can be slipped off again.

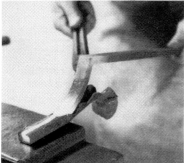

Q

Now weld the two together. First remove the leaf and take a FULL WELDING heat on the bar. Then slip the leaf back, tighten with a hammer and heat both leaf and bar to welding heat. If they were assembled cold the leaf, which is both thinner and more exposed, would burn through before the bar was hot enough. A little silver sand protects the leaf from wasting in the fire.

R

Weld quickly with fairly heavy blows. Do not draw the tip more than necessary or it will have to be upset again in order to weld it on to the next piece.

S

Here, leaf and bar have gone together perfectly, the heat clearly penetrating right through.

CHAPTER 3

Lesson 15   SQUARE BLOCKINGS FOR GATE RAILS

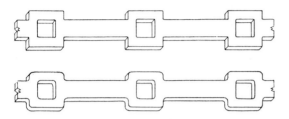

Fig. 30

Square blockings are used when it is desired to reduce the width of the horizontal bars in a gate between the holes where the vertical bars go through them.

When the design calls for sharp square corners the blockings are cut from flat bar giving a look of precision to the work. Where a small radius on each corner is allowable the blockings are punched, upset, drifted and worked up. This gives a more rugged appearance.

Cut-square blockings are either sawn down or cut with a sharp hot set to the required depth. The sides are then cut away with the hot set, and the raw edges levelled up with a square-edged set hammer. The holes are drilled and filed out square. A series of blockings to measure should all be marked out on one bar, allowing a little for draft on the levelling up. Sometimes it is necessary to upset one which has been drawn too much, or vice versa.

A

When the required length of the bar is greater than can be conveniently handled in one piece, make the blockings in two or more lengths and weld these together.

B

To make forged blockings, first upset the bar slightly and slot-punch the hole.

C

Drive out the burr over a bolster.

D

Put a drift in the slot and level up the uneven swelling caused by the punch.

E

Take a NEAR WELDING heat, localise it with water and upset the metal about the slot until the hole becomes round.

F

Level up with a flatter.

G

Place a drift in the hole with great care, so that it is both square and central, and drive it in gradually from each side in turn, but do not drive it right through yet.

H

With the drift still in place, flatten the side slightly.

J

Now drive the drift in a little farther and forge the shoulders to a small radius, between top and bottom fullers.

To finish, level up the bar beyond the shoulders, flatten the sides a trifle more; knock out the drift, flatten the faces again and finally replace the drift and drift out to the correct size.

Fig. 31

CHAPTER 4

MAKING AN ORNAMENTAL GATE

In Part I the making of scrolls was analysed in detail. In Part II these techniques are applied to the making, fitting and assembly of an ornamental gate.

The first stage in making the gate is the framework: the three rails, the front and back stiles, and the vertical bars. First is the making of the bottom rail with its heel; the top rail is similar. The tenons for the centre rail come next and the rail is drilled for the vertical bars during the fitting. The front and back stiles are made in the same way, so only the back stile, with the double running scrolls at the top) has been described. A section has been included on making the latch slot in the front stile. The journal on the back stile is turned or filed during the fitting. When the vertical bars have been tenoned, the framework is ready for fitting and assembly.

The forging of branching scrolls and balls is described next. These techniques are needed for the centre panel and the scroll-work above the top rail. The construction of one of the side panels, which consist of eight scrolls branching from a stem, is described in detail. Once you have mastered these methods you will be able to see for yourself how to make the rest of the scroll-work for this gate. Finally the dog bars, hanging and latch fitting are explained.

The fitting and assembly of the gate is described in Chapter Five.

HEIGHT AND WIDTH ROD

A

The first step in making a gate is to take a clean straight wooden rod and from the drawing transfer the dimensions of the frame accurately on to it, the height on one side and the width on the other.

Do not trust the drawing or print. Even if the draughtsmen were absolutely accurate, the paper may have shrunk or swollen. Find out the measurements and mark the wooden rod with a rule and square.

B

Straightening the Bars
Cut all the bars to length, and take out the twist. Sight through a pair of winding bars made of 'T' iron notched in the middle.

C

To take out the twist with a wrench, grip the bar in a vice and support the end near the wrench with a notched piece of wood, or an adjustable stand.

Remember that it is far easier to untwist a single bar than a whole gate, which you will certainly have to do if the bars you make it from are not straight.

D

Next straighten the bars over the swage hole, or one of the notches in the swage block if the bars are heavy.

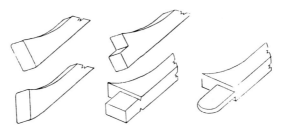

Fig. 32

A

The next job is to upset on the anvil the heel at the hanging end of the bottom bar.

The picture shows the stance and force needed. Take a WELDING heat about 4 inches long and upset by bringing the bar down as hard as you can.

B

Do not allow the bar to buckle very much. Let the striker be ready to give a quick straightening blow and return to the upsetting. A really skilled man can complete the upset in three heats, each shorter than the last. A beginner need not worry if he takes six.

C

Here is the finished upset.

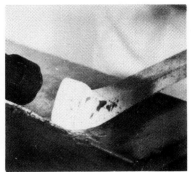

D

Now take another heat, and forge back the corner of the upset to an angle of about 75°.

E

Put a square-edged set hammer just below the edge of the upset. Have your mate lower the haft of his sledge, and strike the near side top edge of the set hammer so that the force of the blow is diagonal to it, and drives the far side bottom edge into the metal. This produces an offset shoulder. Both the angle of the blow and its effect are clearly shown.

F

Hand your hammer to your mate, pick up a side-set, and square up the shoulder with lighter blows.

G

Leave the end a little wider than the bar. There are two reasons for this. First, the metal at the edge will be dragged in somewhat when making the tenon and secondly, a little extra width will be needed for final dressing.

H

Take a butcher and notch both sides of the shoulder. Be careful not to drive too deep, or the tenon will be galled in forging.

J

This is how the job should look now; it is shown cold.

K

Reduce the tenon with a square-edge set hammer over the near edge of the anvil or a square-edge stake (see page 19).

L

Square up the sides of the shoulder with a side set, the striker using a hand hammer as before.

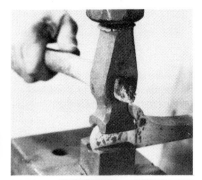

M

Round up the edges of the tenon between swages. This is because a round-ended mortice in the stile is both stronger and easier to make than a square-ended one.

N

Draw out the tip of the heel with a fuller.

O

Finally square up the shoulders with a square-edged set hammer and side set. Test from time to time with a blacksmith's square. By these methods the work can be so accurately finished that little filing will be needed.

P

Here is the finished heel bar. The top rail is made in the same manner.

Fig. 33

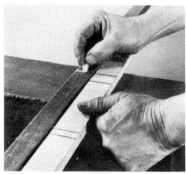

A

Transfer the marks from the wooden rod to the appropriate bars with an engineer's chalk pencil filed to a chisel point.

B

Mark where the shoulder is to come. With the bar on edge cut both shoulders to about three-quarters of their final depth with a hot set.

C

Tap down the burr with the back of the set.

D

Using an anvil saddle or an old piece of plate, cut the sides of the tenon, one from each side of the bar.

E

On the last of the heat, chisel off the sharp corner left on opposite edges of the tenon by the previous two cuts.

F

Swage the tenon.

G

Square the shoulder with a monkey tool.

H

Trim the corners of the bar so that the end of the rail is as wide as the stile.

J

For measuring shoulders, make a tenon gauge from any light bar near to hand, with one end bent and forked like this.

K

Mark off the length of the tenon gauge from the height and width rod, and cut it off square. Then mark off the other end of the rail with this gauge.

L

Form the second tenon on the rail, and monkey it up to length, measuring with the gauge. Allow about 1/16th of an inch for shrinkage and fitting with mild steel, slightly more with iron.

Fig. 34

This stile is finished with a double running scroll at the top. A few inches of the bar are forged out, and the double running scrolls welded to it.

A

Set in with a cheese fuller.

B

Draw out to a taper, with the end the same size as the scroll bar which is to be welded to it.

C

Form the smaller scroll from a suitable bar. Mark the bar at any convenient point near the scroll and mark the plate (see page 66) at the same point.

Also mark the plate at the point where the two scrolls join.

D

Measure the distance between the marks with string. Measure off this distance from the first mark on the scroll bar and centre-punch.

Estimate how much metal to allow for the weld which is to form the tip of the double running scrolls. Leave too much rather than too little.

E

Cut off enough metal to form the scroll between the bolt end and the stile.

Clamp the two bars together with tongs and weld them lightly. Remove the tongs. Then take a FULL WELDING heat and weld to the punch mark. Draw the weld out to a fishtail and form the bolt-ended scroll, as described on page 25.

F

Here the bolt is being finally dressed.

G

Form the scrolls with hammer, horns and wrench. Here one scroll is being rectified on the chalked metal plate.

H

Set a part of the scroll correctly to the drawing on the plate. Make a centre-punch mark on the scroll bar where the straight bar leaves the scroll line on the drawing, and a corresponding mark on the drawing. Measure on the drawing the distance with string from the shoulder on the back stile to the mark, and set the dividers to this measurement. Check and cut off the surplus metal, upset, scarf and weld.

J

Draw to the correct length and check with dividers.

K

Finally, set the scroll-work to the correct curve with two wrenches, aided by the horns if necessary.

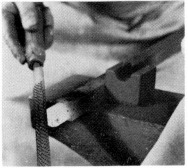

L

Mark out the back stile from the height and width rod, and cut it to length. Butcher and swage the turning pin; trim to length and dress the round end of the pin with a hot rasp.

Lesson 20   **THE LATCH SLOT**

A

Punch the hole with a slot punch and drive a drift part way in, over a bolster.

B

Dress up the edges with a flatter, as shown here. Knock out the drift and flatten the faces. Drive the drift right in and work up with a flatter to a good finish.

Lesson 21   **VERTICAL BARS**

A

The tenons of the vertical bars are butchered as shown here and swaged up.

Mark out the other end, forge the second tenon and monkey to length as shown on page 63 (F and G).

68

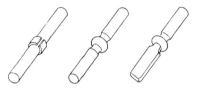

Fig. 35

A

Balls are a common embellishment of wrought ironwork. They are made by welding a collar to the rod in specially-shaped tools.

First make a collar to fit the round bar, but leave a slight gap. Close it, cold, on to the bar at a NEAR WELDING heat. Otherwise the collar would be burnt before the welding heat penetrated to the centre of the bar.

B

Return collar and bar to the fire, and while raising a welding heat, put a few drops of oil in the bottom tool.

C

At a FULL WELDING heat, place the collar between the tools and weld quickly. The oil releases the ball from the bottom tool, where it might otherwise stick, and helps to give a smooth finish.

D

Here is the finished ball in the bottom tool. Note that it was made so quickly that it has scarcely cooled yet.

E

Flatten the bar behind the ball.

F

The weld shown here is required in the making of the centre panel.

Fig. 36

A

To make the top of the centre panel, swage down the metal on one side of the ball to a ⅜-inch pin, and flatten the bar on the other side. Measure the scrolls as described on page 66 D and grip them to the flattened bar with the tongs.

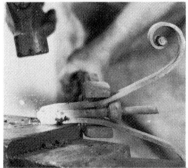

B

Take a LIGHT WELDING heat, and weld the ends together.

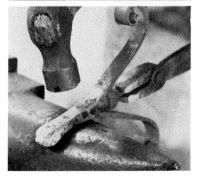

C

Now grip the pin in a suitable pair of tongs. Take a FULL WELDING heat and weld securely. Draw out the weld.

D

Weld the second pair of scrolls at the position marked on the drawing.

E

Weld on the lower ball, and draw out the weld to length. Here the length of the finished weld is being checked with dividers.

F

Finish the two pairs of scrolls with scroll tool and wrench.

G

Set them to the chalked plate.

First transfer the whole panel of scrolls on to one or more convenient pieces of plate (see page 12). Decide how best to build up the panel, where to place the welds, and how far to roll the scrolls before welding them.

The panel for this gate is made up in five parts:

1. The top scroll to K.

2. The next two scrolls and the piece between them to L.

3. The two middle scrolls and the piece between them to M.

4. The lower pair of scrolls and the piece to N.

5. The bottom scroll.

Fig. 37

First make the three pairs of small scrolls between K and N, as follows:

Mark the drawing at the points where the scrolls branch from the pieces which form the stem. Measure the length from these marks to the snub in each case. Forge the fishtail snubs on each of the six scrolls.

Mark off the scroll bars to length, centre-punch them, add 1½ inches for the weld and cut off.

Cut the three intermediate pieces which form the stem, centre-punch marking them at either end and making the same allowances for the welds.

A

Here is a scroll with the tip rolled, and an intermediate piece being punched.

B

Next grip one of the scroll bars and the connecting piece in tongs as on page 66, weld up to the centre-punch marks and draw out the weld.

Roll the scroll a little more so that it clears the far end of the connecting piece.

C

Now weld on the second scroll in the same way.

D

Lay the piece in position on the plate and mark where the end of the stem comes.

Roll the beginning of the top scroll. Lay it in place on the plate and mark both the scroll and the plate with chalk where the bar becomes straight.

E

Measure the distance between these two marks with string. Then mark off this length on the top scroll, add an allowance for welding and cut off.

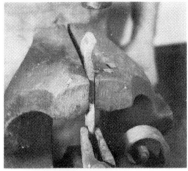

F

Upset the end of the partly-formed top scroll in the vice.

G

The end is bound to buckle. Straighten it with a light hammer against your hand hammer.

H

Weld the top scroll to the pair of partly-formed scrolls which have already been welded to their connecting piece and then set to the chalk drawing.

Weld on the next section of the stem to which a pair of scrolls have already been welded.

J

Check the length with dividers before rolling up each pair of scrolls. It is essential that this measurement is correct.

The remaining pair of scrolls and the bottom scroll are welded on in the same way and set to the drawing.

K

Much of the rolling can be done on a scroll tool.

L

But some parts will need horns and wrench.

M

A file tang is useful in setting the scrolls accurately near the branch of the weld.

Lesson 25   **DOG BARS**

A

Fig. 38

The partly rolled dog bar scrolls are welded to the short centre spike as on page 71 and finished on the scroll tool. The drawing shows the complete scrolls.

Lesson 26   **HANGING FOR THE GATE**

The gate is supported by the turning pin at the bottom of the back stile (see page 67) which turns in an iron socket let into the ground.

The top is held by a bearing. A bar is built into the gate post, and fitted at the end with a strap which passes round a 'journal', a part of the back stile rounded for this purpose and described under fitting on page 84.

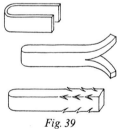

Fig. 39

To make the bearing, the bar is forged with a convex end, the strap fitted to it, and a hole drilled through to make a machined bearing (see Lesson 32 B).

B

Cut a piece of bar to length and trim off the corners with a hot set.

C

Swage the end to a half circle.

D

Heat the other end of the bar and if it is to be set in stone with lead, rag it with a hot set, as shown here –

E

– or if it is to be set in brickwork, split it and splay it out.

F

Next make the strap. As a flat bar will hollow if it is bent to a small radius, it should be dished before bending. Use a large bottom swage and a cheese fuller.

G

Now bend the strap, rounded side outwards, over the bick. Instead of hollowing, the strap will have a pleasantly rounded look, and fit closely to the journal.

H

Dress the strap down solidly on to the bar, and put aside until the fitting.

THE LATCH PIVOT

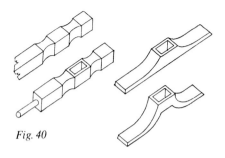

Fig. 40

A

Take a piece of metal as big as the largest section of the latch pivot, and fuller it like this.

B

Punch and drift the slot and cut off.

Draw down the ends with cheese fuller and flatter.

C

Make the sharp bend in the vice with the aid of a cheese fuller; note the sheet metal clamp in the vice to prevent galling.

THE LATCH

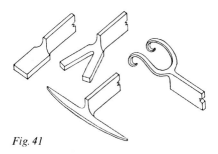

Fig. 41

A

Flatten the end of the latch bar offset and split it as shown.

B

Turn the two ends to a right angle, and draw each to a long tapering point of oval section.

C

Curve these ends and scroll the tops like this.

Assembling the gate

CHAPTER 5

FITTING AND ASSEMBLING THE GATE

When all the parts have been forged, the ate is ready to be fitted together and finally assembled.

This includes making the journal on the back stile, slotting and fitting the heel bars; drilling the front stile and the top and bottom rails; making the square holes in the middle rail, fitting the uprights and riveting them all together; fitting and fixing the scroll-work and the latch and completing the hinge piece.

When the final assembly is complete and any roughnesses have been dressed off, the job is ready for painting.

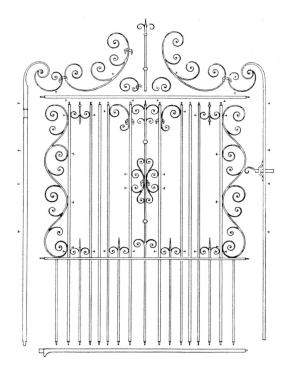

Fig. 42

A

Chalk the back stile and mark the journal from the height and width rod.

B

Cut into the corners at intervals of about ¼ inch with a hack-saw and cut them away with a cold chisel, so that the bar becomes eight sided. File off all the corners, and the journal is completely round.

It is of course quicker to turn it on a lathe, if you have a suitable one.

C

Slotting the Back and Front Stiles
Next make the slot hole. Mark out and drill two holes. Plug them with small pieces of round bar. Centre-punch mark a central hole between them and drill it out with a pilot drill about $\frac{3}{16}$ inch.

D

Drill out to the full size, and knock out what is left of the plugs. The slot can be filed out to size quite easily.

Before drilling the back stile for the top rail, or the front stile, make sure that all the uprights are long enough. They should have been monkeyed to $\frac{1}{16}$ inch, over length. The bars will then only require the minimum of fitting.

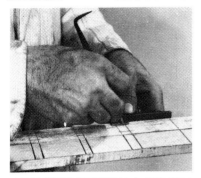

A

Mark out the position of the uprights on the middle rail from the height and width rod with a square as on page 57. The ends of the holes can be conveniently marked with a combination square used as a depth gauge.

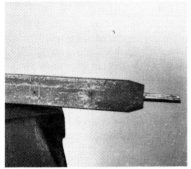

B

Centre-punch mark the corners of the holes, and suitable places for drilling and plugging as on the opposite page.

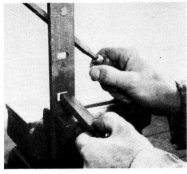

C

Here are three stages of making the slot holes for the uprights in the centre rail: a plug in place, drilling completed and plugs removed, and a finished hole after filing.

D

Fitting the Heel Bar

Fit the heel bar to the back stile: this is the vital joint. The soundness of the assembled gate will depend largely on how carefully this is done.

As most gates drop a little when hung, do not set the rail square with the stile, but give it a rise of $\frac{1}{16}$ inch for each foot the gate is wide. The tapering gap can be clearly seen between the square and the rail.

The heel should be about $\frac{1}{16}$ inch wider than the stile (see page 59). Do not reduce it yet.

A

When the heel bars have been fitted, and the remaining round holes for the tenons drilled in the top and bottom rails and the front stile, the frame can be put together, but not riveted up yet.

For this you need two trestles. See that they are accurately aligned.

B

Put the two stiles and the top and bottom rails together on the trestles.

C

Try one of the vertical bars between the top and bottom rails, and file to length, if necessary.

D

Now file all the uprights to length, checking with a tenon gauge.

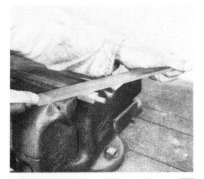

E

The ends of the vertical bars may still be upset slightly from the monkeying. If this is so, dress them with a file. Otherwise you will have to make the holes in the middle rail too big, and once the thickened end has passed through, the bar will be loose.

F

Turn the frame so that the rails lie on the trestles, and remove the stiles. Lay the middle rail in position, thread the vertical bars through it, and enter the tenons into the top and bottom rails.

G

Fit the centre panel bar. On this bar there is one ball above and another below the top rail: these are screwed together. The ball below the rail has a pin forged on it which is screwed; the ball above is drilled and tapped. If they do not tighten up at the right point, file a little off the top ball.

H

Replace the stiles, and see that all the joints fit comfortably.

Check the squareness of the whole gate, by the diagonals this time, not forgetting the slight rise mentioned on page 85 D.

87

J

As the main frame should not be riveted before the scroll-work is fitted and the fixing holes for it drilled, the frame should be clamped up tightly, to hold tenons and bars together while the scroll-work is being fitted.

Use a pair of joiner's clamps if you have them; if not, make one or more cramp bars as shown from any material to hand, and force them on.

K

Now do the final cold setting of the scroll-work. Scroll-work must never be sprung in or it will certainly distort the frame; any adjustment should be made with the horns and scroll wrench.

L

Mark the bars where the scrolls are to be screwed to them. Dismantle the gate once more, and drill and countersink the holes.

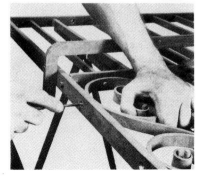

M

Re-assemble the gate, and mark the scrolls through the holes in the bars with a scriber.

A

Centre punch the scrolls, drill them tapping size over a bar bolted to the drilling machine table, and tap them.

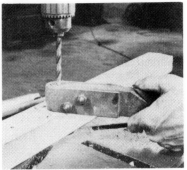

B

Fitting the Hanging

Take the strap and bar, drill through both and fit two bolts.

Mark the centre of the circle where the journal is to come, in such a position that the drill will cut slightly into the strap.

Drill a pilot hole right through.

C

Then drill out to full size.

D

Riveting the Gate

Here is a single-handed method of riveting upright bars.

Rivet the uprights first. Place a strip of metal under the bar on which you are working to raise the other tenons clear of the anvil face. Rivet the end bars top and bottom first, to keep the frame under control.

E

Rivet each tenon with the peen of the hammer until it grips firmly.

F

Sight the gate for alignment.

G

Check the squareness roughly from time to time, not forgetting the rise.

H

While your mate holds a sledge hammer against the far end of the bottom rail–

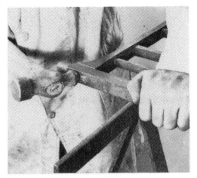

J

– rivet up the tenon of the heel.

K

The heel was left a trifle wider than the back stile (see pages 59 and 85).

Now, with a sledge hammer below it, burr it up flush with the peen of the hammer. This ensures firm contact at the outside edges and tightens the joint.

Rivet and finish the top heel in the same way.

L

The small spiked ornaments under the top rail are difficult to rivet up as the spike is not stout enough to resist the blows. Make a holding-up dolly as shown, by drilling and countersinking the end of a heavy bar. This will fit over the spike and rest against the ball. Alternatively use a lead or lead and antimony block held against the spike.

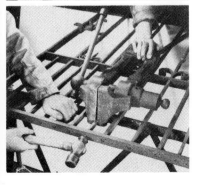

M

Use a leg vice for riveting dog bars.

A

Secure the scrolls with countersunk screws. Forge a special screwdriver as shown, having one end in line with, and the other at right angle to, the shank.

B

Either saw them off with a hack-saw –

C

or cut off the screw ends with a sharp cold chisel –

D

– and dress them up with a half-round file.

The remaining scroll-work is fitted in the same way, except that seven collars are used. Heat the collars and place them on as described on page 41.

The gate is now complete except for painting.

PART III

CHAPTER 6

PAINTING WROUGHT IRONWORK

All ironwork must be protected against corrosion. In particular gates, railings and signs which stand in the open must be painted most carefully if they are not to rust, especially as many smiths now work in mild steel which rusts more easily than pure iron. A thorough protection may be expensive; but a customer who accepts a poor finish to save money will find that he must either pay more in maintenance, or see the work ruined.

PRIMING

Before applying a priming coat, remove all the loose scale produced in forging by chipping, wire brushing, and polishing with emery-cloth. Pieces of loose scale left on the iron will sooner or later crack the skin of the paint. Mill scale, which is not so loose, need only be removed for a bright finish. Any rust on the iron must also be removed. There are several reliable brands of derusting fluid on the market. The main agent in most of these is phosphoric acid, which converts whatever rust remains after wire brushing into a grey deposit, iron phosphite. The metal can usually then be painted without further treatment, but the paint makers' directions should be carefully followed.

When the iron has been thoroughly cleaned of scale, and the rust treated, it is ready for a priming coat. Red lead and red oxide are often recommended, and will protect the metal well, provided they are of good quality and applied at the right consistency. A mixture of red lead powder, japan gold size and *pure* turpentine makes a reliable primer which, although it dries quickly, will not crack. Properly mixed, it dries matt and provides an excellent key to the next coat. Add the japan gold size to the red lead powder until the mixture becomes as stiff as a very thick cream. Then thin it gradually with pure turpentine to the consistency of thin cream. Leave the mixture to stand for one or two hours. Stir vigorously and add a little more turpentine if necessary. Stirring frequently, apply the primer so thinly that the colour of the metal shows through the priming coat. The paint should be dry and ready for a second coat in three to six hours, according to the temperature of the workshop.

This primer will not round off or blot out sharp detail, and will not readily chip. No primer will be efficient if it is applied thickly, for it will bridge the small pits and pores of the metal without entering them. Two thin coats are much better than one thick coat. The first coat at least must be thin enough to fill the pores and crevices, even though it may not cover the whole surface of the metal opaquely.

In the past few years several zinc-based paints have been introduced into the market. Research into these paints was carried out at the Cavendish Laboratory in Cambridge, and stringent tests have proved them to provide a better protection against rust than most other priming paints. These zinc paints can be applied over rusty metal, provided the rust is not loose, because the chemical reaction of the particles of zinc in the paint against the metal reduces the rust and prevents it developing beneath the paint. For this reason zinc paints are generally described as 'cold galvanizing' paints. Most oil-bound finishing paints can be applied over them. There is at least one brand of zinc paint which can be applied to wet metal.

Zinc paints must not be confused with aluminium paints. Some aluminium paints are efficient primers, provided they are made from aluminium leafing paste, but unlike zinc paints they will have no beneficial chemical action on any rust over which they are painted. Take the maker's or dealer's advice before choosing a paint to cover the aluminium primer, because some undercoat and finishing paints are mixed with a base which softens the medium which the aluminium is mixed with. If the next coat of paint picks up the aluminium the job will be spoiled, and you will have to start from the bare metal; this might involve the expense of having the work sand blasted.

UNDERCOAT

For the undercoat a flat lead paint, either prepared locally by a good ironmonger or a proprietary brand, is usually best. But take care that your primer, undercoat and topcoat paints suit each other chemically. Some prefer a grey undercoat to black, so as to see more easily the parts which have still to be painted when the black topcoat is applied.

TOPCOAT

Most people prefer an egg-shell black finish, and there are several reliable black lacquers with this finish on the market. Many experienced smiths consider berlin black ideal. Others finish with a gloss paint which is unpleasantly shiny when fresh but soon weathers.

Whatever paint you use, apply it thinly, so that none of the delicacy of the work is spoiled. Thick leathery paint not only blunts the details, it does not protect the work so well. It will crack and blister in summer, and damp will enter through the cracks, rusting the metal beneath the paint.

Wrought ironwork used in the house, such as pokers, tongs and firescreens, must be as thoroughly protected as outdoor work. For although indoor work will not rust so quickly, damp, dust, and the corrosion of constant handling can wear a handsome piece into scrap iron. Painting protects the work, gives it an even colour, and makes it easier to clean. Alternatively, indoor work is sometimes finished armour bright.

ARMOUR BRIGHT

If a piece is to be finished armour bright, it must be forged with much greater care and skill than would be needed if it were to be painted. The fire must be clean, and the hammer and anvil must have smooth faces, free of dirt and scale. The metal must also be cleaned of dirt and scale with an old file before it is worked. Scrape the iron but make no attempt to file it up, as this would give an uninteresting machine-like surface. Any resulting marks on the cleaned surface can be removed by draw filing. Clumsy or unnecessary filing are the commonest faults of armour-bright work.

Complete all the fitting before you polish the work. If you have several pieces to finish armour bright, you will save much time and labour by making a 'pickle' of five parts water to one of sulphuric acid, which will remove all the scale, including mill scale, from the iron. The pickle must be mixed in a bath which will not be corroded by the acid. It is important that the water is put *first* into the bath and the acid added to it; otherwise there is danger of an explosion due to the rapid generation of steam. The piece should be taken out of the acid bath and brushed with steel wire once or twice until all the scale is gone. Rinse it first in cold water, then in sulphate of ammonia to destroy any acid that remains, and then in cold water again. Finally dry it in a box of sawdust.

The piece is now ready for polishing. It is much less laborious to use a power polishing bob, but if you work by hand use emery-cloth which, although more expensive than emery-paper, lasts much longer. After polishing, take care not to handle the work with bare hands if it is to be lacquered. Use a duster or rag, otherwise the fingerprints will show through the lacquer.

Armour-bright work is often lacquered to preserve the finish. Lacquering will last much longer if properly stoved by a professional lacquerer. Cold lacquering is practicable, but the temperature must be at least sixty-five degrees fahrenheit, and the atmosphere free from dust.

In this chapter specific brands of paint have not been recommended, because good new paints are continually appearing, and there are many brands which the Commission has not been able to test. But the techniques described, used with good paints, will afford a tough protection and clean finish. They take time, and may therefore prove expensive. But unless wrought ironwork is properly painted by someone (the smith himself may not always be the right man), it will not last, as good ironwork deserves to last, for the enjoyment of many generations.

OTHER BOOKS PUBLISHED BY
THE RURAL DEVELOPMENT COMMISSION
141 Castle Street, Salisbury, Wiltshire SP1 3TP

The Blacksmith's Craft
ISBN 1 869964 14 4

Blacksmith's Manual illustrated by J. W. Lillico
ISBN 1 869964 21 7

Catalogue of Drawings Jo; Wrought Ironwork
ISBN 0 854070 28 1

Catalogue of Drawings Wrought Ironwork Gates
ISBN 1 869964 22 5

Catalogue of Drawings Weathervanes
ISBN 1 869964 28 4

Decorative Ironwork
ISBN 0 854070 12 5

Metals for Engineering Craftsmen
ISBN 0 854070 27 3

Today the Commission, which has always been involved in promoting traditional rural crafts in England, also offers training.

Forgework courses are set at various levels and cover: general smithing techniques, scroll work, fitting and framework, power hammering and toolmaking, art metalwork, block repoussé, gilding and decorative effects. For details please contact the Training Section in Salisbury.